VALIDATION OF EQUIPMENT/ INSTRUEMENT

Definition:

- Validation is performing to ensure that a system or piece of equipment is reliable meet requirement of CGMP and that the performance is reproducible.

Another definition given by FDA 1983 is:

- A document program which precedes a high degree of assurance that a specific process will consistently produce a product meeting its 3 determined specification and quality attributes.

A more simple definition is:

- A feasible way to attained assurance that the method is workable.

In short we can define validation:

- The documented act of proving that any procedure work correcting and lead to the expected result.

Steps of Validation

Equipment Qualification:

Qualification is required for all pieces of equipment that are used to manufacturing product, supply utilities and ancillary equipment (that supports main equipment pieces).

The qualification process consists of three mandatory parts.
- i- Installation Qualification
- ii- Operational Qualification
- iii- Performance Qualification

i- Installation Qualification:

It is the process of achieving documented evidence, which involves the equipment installation phase and is designed to compare the equipment to the original design and to ensure that the utilities and ancillary system connected to the equipment are in accordance with the manufacturer's specification.

ii- Operational Qualification:

It is the process of achieving documented evidence, which involves the test each function of the equipment and ancillary system and to ensure that the equipment is properly functioned and operated throughout the recommended operating ranges.

iii- Performance Qualification:

It is the process of achieving documented evidence, which involves the performance of equipment to meet predefined appropriate standards.

There are three type of PQ validation.
1. Retrospective: This qualification is applicable on previous manufactured batches. The result and different data of minimum 11 batches are collected and on the basis of their result.
2. Prospective: This qualification is performed on different three full scale batches before manufacturing the product.
3. Concurrent: This type of qualification is performed on three running batches.

Before purchasing any equipment following documentation are required.

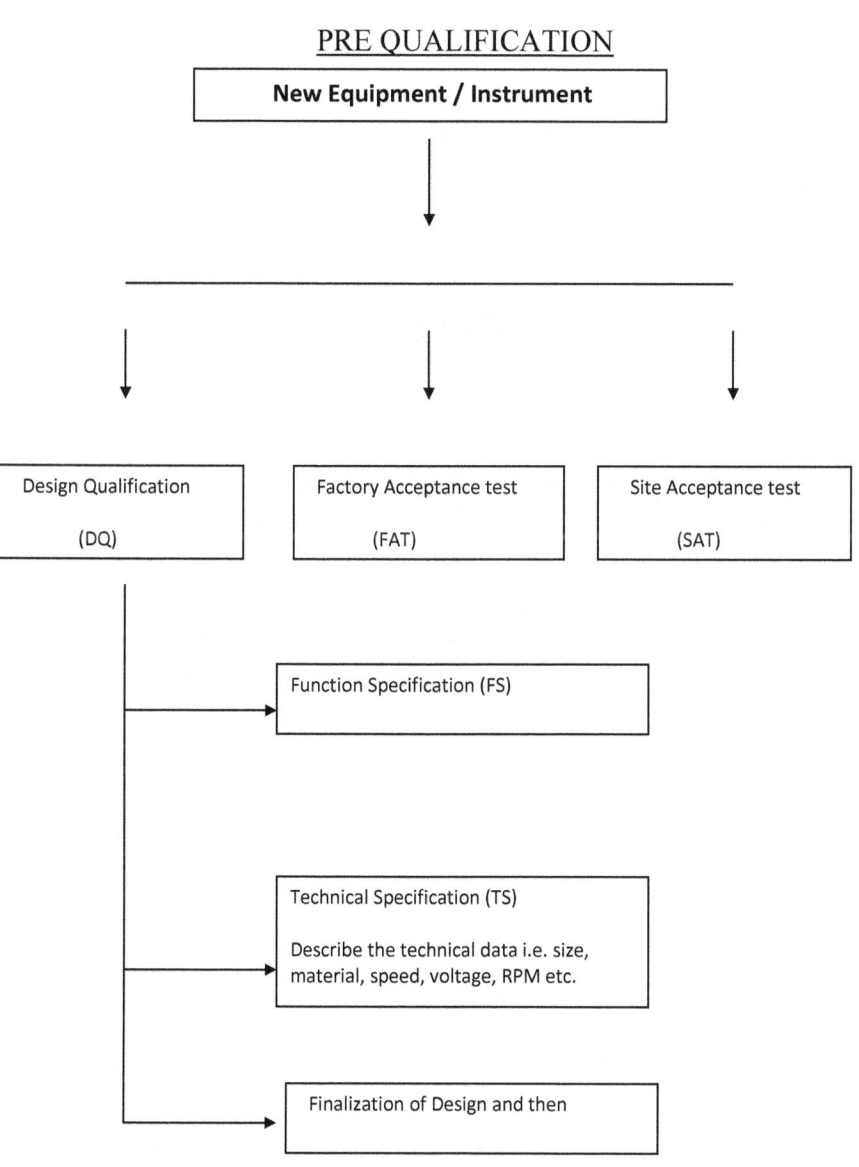

POST QUALIFICATION

<u>After purchasing of equipment following validation plan should be followed</u>

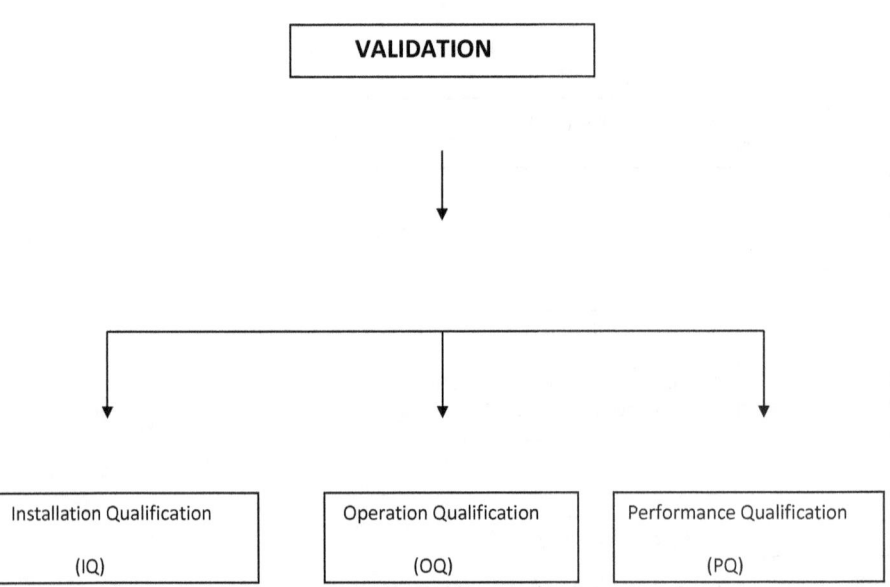

ABC PHARMACEUTICALS		Quality Control Department	
TITLE : Validation of Equipments		Document # AB-VDI-028	
Subject: Validation of Tablet Coating Pan		S.O.P # AB- VDCP-001	
Page # ……….	Review # 000	Equipment # ABPD –T/50 & 60-A to 70-I & T /79	
Effective Date: ………………		Review Date :- If any change in equipment and method	

Scope: To validate the equipment of film coating, sugar coating & enteric coating pan for accurate result.

Purpose: Ensure that the supply utilities and ancillary system connected to the equipment are in accordance with the manufacturer's specification.

Tablet Coating Pan:-

The coating pan system consists of a circular stainless steel pan 304 mounted somewhat angularly on a stand. The pan is fitted with integral baffle which is rotated on its horizontal axis by a motor .Heated air is directed in to the pan through a blower and onto the exhaust system that is duct, positioned through the front of the pan. Coating solution is applied to the tablets by ladling or spraying the material onto the rotating tablet bed. Spraying can significantly reduce drying time between solution application in sugar coating process and allow for continuous applications of the solution in film coating.

INSTALATION QUALIFICATION

Equipment Identification:

Required Information	As-found conditions	
	Film coating pan	Sugar &Enteric coating pan
Manufacturer Purchase Order No. Model No.		

Serial No.	FD&C	FD&C
Equipment NO.	Nil	Nil
	Nil	Nil
	Nil	Nil
Location	ABPD –T /79	ABPD-T /70
Diameter		&
Capacity		ABPD-T / 70-A — 70-I
	Coating Dpt.	Coating Dpt.
	36 inch	72 inch
	30 Kg	60 Kg

UTILITIES:-

Utility	Specified	Measured Result	Acceptable (Y /N)
Volts	Coating pan = 380 volt ±10%	395 volt	Yes
Amps	Coating pan motor =8.8	8.7	Yes
	Supply blower = 3.5	3.7	Yes

	Exhaust blower = 5.4	5.4	Yes
Compressed Air	Thermostatically controlled heating system	Complies	Yes

Instrument Used:

Test Instrument	Identification No.	Calibration due date
Multimeter	Nil	Nil

Volt Calculation:

Coating pan motor volt specification = 380 V ± 10 %

±10 % OF 380 = ± 38

380 + 38 = 418

380 - 38 = 342

The measured volts of 395 fall within the ± 10%

Amp Calculation:

Circuit rating = 20 A

Coating pan motor = 8.7 A

Supply blower motor = 3.7 A

Exhaust blower motor = 5.4 A

Equipment current draw = Coating pan + Supply blower + Exhaust blower

 = 8.8 + 3.7 + 5.4

 Total = 17.9 A

The circuit amp rating observed 17.9 A fall within the range of 20 A.

MAJOR COMPONENTS SPECIFICATION:

S.NO.	Components	As-found Conditions
1,	Coating pan motor	Manufacture : Siemens Volts : 381volt – 391 volt Ampere : 8.8 Phases : 3 Pan's rpm : 50 Hz Hp : 5 hp Motor's rpm : 1435
2,	Supply blower motor	Manufacturer : Siemens Volts : 381 volt – 391 volt Ampere : 3.7 Phases : 3 Cycles : 50 Hz Hp : 3 Rpm : 1420
3,	Exhaust blower motor	Manufacturer : Siemens Volts : 381 – 391 volt Ampere : 5.4 Phases : 3 Cycles : 50 Hz Hp : 5 Rpm : 1420

4,	Spray gun	Manufacturer : China
		Type : Cyclic
		Nozzle dia : 0.2mm -0.5mm

MATERIAL USED:

Record material of each component that contacts the products.

Component	Material
Coating Pan	Stainless Steel 304

Lubricants:

Where used	Type	Manufacturer	Product Contact Yes / No
Motor Bearing	Lubricant Oil 30 /40	Any Oil Co.	No

Equipment Safety Features:

- Always ensure that the switch is off the main of coating pan, hot air blowers and exhaust system.
- Blower Temperature should be monitored whether the pan supply temperature controller operates according to the indicated temperature.
- Exhaust system should be monitored whether it is working proper or not.

OPERATIONAL QUALIFICATION

An OQ evaluation should establish that the equipment can operate within specified tolerances and limits .The tablet coating pan will be validated for its operating ability, The information required for the OQ evaluation is calibration of the instrument used to control the tablets coating pan, equipment control function (switches and push buttons) and equipment operation (pan motor rotation direction, pan motor speed, pan supply temperature control, inlet air volume).

Calibration Requirement:

Verify that the equipment have been logged into the calibration system. Record information to control the coating pan.

- Count the number of rotation of tablet coating pan
- Check pan supply temperature.
- Check the blower temperature.
- Check the blower air volume.

Equipment Control Functions:

Test Function	Expected Results	Acceptable (Yes / No)
Main power switch	At OFF position no current or power in main panel.	Yes
	At ON position power supply ON.	Yes
Black push button for coating pan	When the pans start, black push button pressed, the pan starts and rotates continuously	Yes
Red push button for coating pan.	When the pans stop, red push button is pressed, the pan stops rotating.	Yes
Green push button for blower	When the blower start, green push button is pressed. The blower starts.	Yes
Red push button for blower	When the blower stop, red push button is pressed, the blower stops.	Yes
Green push button for Exhaust system	When the exhaust system start, green push button is pressed, the exhaust start.	Yes
Red push button for	When the exhaust system stop, red push button is	Yes

exhaust' system	pressed, the exhaust stop.	
Temperature controller	Through membrane push ON square located on main panel, temperature function display. Set it on required temperature. Red light indicates that required temperature has been achieved.	Yes

Equipment Operation:

Tablet Coating pan Rotation Direction Test

It is to verify that the tablet coating pan rotates in the proper direction. The tablet coating pan will be operated with the pan empty. Press the start push button and observe the direction of rotation of the tablet and record the result.

Item	Expected Results	Result	Acceptable (Yes/No)
Tablet coating pan rotation direction	Rotation should be clockwise	Clockwise rotation was observed	Yes

Tablet Coating Pan Speed Test Result

Item	Specified speed (rpm ± 2)	Measured speed (rpm)	Acceptable (Yes /No)
Pan speed (ABPD-T /70 & ABPD-T / 70-A — 70-I)	15	14	Yes

Film coating pan Speed Test Result

Item	Specified Speed (rpm ± 1)	Measured Speed (rpm)	Acceptable (Yes / No)
Pan speed (ABPD –T /79)	10	10	Yes

Instrument Used:-

Test Instrument	Identification no.
Tachometer	ABLI / 64
Stop Watch	ABLI / 73

Pan Supply Temperature Control Test

It is to verify that the pan supply temperature controller operates according to the indicated temperature. Check the controller temperature and record the result. The calibration record will be attached.

Pan Blower Temperature Test

Item	Specified Temperature	Observed Temperature	Acceptable (Yes /No)
Blower Temperature	50°C – 70°C	55°C – 65°C	Yes

Instrument Used:-

Test Instrument	Identification no.
Electronic Thermometer with Stainless steel probe	ISLI / 51

Pan Supply Air flow control test (Sugar & Enteric Coated)

Item	Specified Air	Observed Air	Acceptable

	Volume(CFM)	Volume	(Yes /No)
Blower Air Volume	800 ---1300	1059	Yes

Film Coating Pan Air Flow Control Test

Item	Specified Air Volume(CFM)	Observed Air Volume	Acceptable (Yes /No)
Blower Air Volume	2000---2200	2100	Yes

Instrument Used:-

Test Instrument	Identification no.
Anemometer	ABLI / 63

Standard Operating Procedures

1. Cleaning Procedure of Coating Pan
2. SOP of Film coating, Sugar coating and Enteric coating Tablet.

Standard Operating Procedure for Cleaning of Coating pan:-

As soon as the coating operation is over, remove the coated tablets in suitable containers. Remove the dry powder from the pan, if any.

Pour approx. 20 liters of hot water (80°C – 85 °C) into the pan and allow the pan to rotate for about 10 minutes .Remove this turbid water using a plastic mug and transfer it into a plastic bucket. Clean the inner surface of the pan with angular wooden brush having nylon teeth.

Place about 500 g of Sodium Hydroxide pellets into the pan and gradually pour about 20 liters of hot water in to the pan, taking care that the eyes are well protected by goggles and full size PVC gloves are worn to cover arms. Allow the pan to rotate for 10 minutes. Remove this alkaline turbid water.

Pour about 10 liters of lukewarm water (40-45) into the pan. Allow the pan to rotate and continue to clean the inner surface of the pan with the angular wooden brush having nylon bristles. Remove the water and check that the inner surface of the pan is free from any powder or foreign particles.

Sampling procedure of Tablet from Coating Pan

Collect ten tablets from three different places (right, left and centre) of each coating pan and mix them thoroughly.

Draw the tablets from mix sample using the following formula.

$$\sqrt{n} + 1$$

Where n is the number of tablets collected. Start the testing according to standard operating procedure.

PERFORMANCE TEST RESULT

ABC Film Coated Tablet:

Test Applied	Limit	Result			Remarks
		390708	390409	151109	
Appearance	White film coated tablet	Complies	Complies	Complies	Satisfactory
Uniformity of weight	± 5 %	Complies	Complies	Complies	"
D.T	Max. 30 mts	10 – 12	29 – 30	12 -- 15	"
Dissolution	NLT 80 %	91.3 %	95.17 %	98.85%	"
Assay: Ciprofloxacin HCL 500mg / tablet	90 % ----- 110 %	102.90%	99.01%	100.05 %	"

ABC SUGAR COATED TABLET

Test Applied	Limit	Result			Remarks
		151008	270209	080809	
Appearance	Blue Sugar Coated Tablet	Complies	Complies	Complies	Satisfactory
Uniformity of weight	± 10 %	Complies	Complies	Complies	"
D.T	Max. 1 hour	12 –14 mts	12 –15 mts	15 –20 mts	"
Dissolution	NLT 70 %	97.57 %	99.77 %	98.16 %	"
Assay: Promethazine HCL 10 mg / tablet	92.5 % ---107.5 %	98.95 %	98.78%	102.09 %	"

ABC ENTERIC COATED TABLET

Test Applied	Limit	Result			Remarks
		350808	110908	110309	
Appearance	Yellow Enteric Coated Tablet	Complies	Complies	Complies	Satisfactory
Uniformity of weight	± 10 %	Complies	Complies	Complies	"
D.T (In HCL)	No released of any ingredient in 2 hours without disc.	Complies	Complies	Complies	"
D.T (In PO4 buffer)	Max. 1 hour with disc.	15 – 20mts	10 – 15 mts	6 – 8 mts	"
Assay:Bisacodyl 5.0mg / tablet	95 % ---105 %	98.61 %	98.81 %	99.85 %	"

CONCLUSION:

When the equipment run according to specified condition by observing the result of three batches of sugar coated, three batches of film coated &three batches of enteric coated tablet. It is found that all test result &procedure of tablets are lying with in limits.

So this is concluded that the equipment &all procedures are validated.

Prepared by:	Checked by:	Approved by:.......................

Following Documents will be attached as per requirement.

TITLE : Standard Operating Procedure (S.O.P)		Tablet Section	
ABC PHARMACEUTICALS		Production Department	
Subject :Outline Of Film Coating		S.O.P # ABIPD-SOP-FC-035	
Effective Date :	Review Date :	Page No	Rev: 000

SCOPE: This procedure applies for general film coating of tablet.

PURPOSE: To provide a method of film coating tablet.

COATING EQUIPMENT:

1. S.S 304 Coating Pan
2. S.S Spray gun
3. Silverson Mixer

COATING MATERIAL:

S.NO	Item	Quantity

ABC PHARMACEUTICALS		Production Department	
TITLE : Standard Operating Procedure (S.O.P)		**Tablet Section**	
Subject :Outline Of Sugar Coating		**S.O.P # ABIPD-SOP-SC-002**	
Effective Date : 15.07.09	**Review Date : 15.07.12**	**Page No. 1 of 1**	**Rev: 000**

SCOPE: This procedure applies for general Sugar coating of tablet.

PURPOSE: To provide a method of Sugar coating tablet.

COATING EQUIPMENT :

1. S.S 304 Coating Pan
2. Silverson Mixer

SEALING STAGE

S.NO	Item	Quantity
1.	Sugar	Q.S As per requisition
1.	Eudragit L100	Q.S As per requisition
2.	I.P.A	"
3.	Talcum	"
4.	P.E.G	"
5.	Food color	" (As applicable)
6.	Titanium Dioxide	"

PROCEDURE:

1. Dissolve Eudragit L 100 in IPA solution and keep it over night.
2. Next day add Titanium dioxide, Talcum, Food color (as applicable) & PEG.Mix with silverson mixer to make a homogenous solution. Coating material solution is ready to use.
3. Use S.S Spray gun nozzle size 0.2mm & apply solution, required quantity with continuous straining.
4. Rotate the pan with slow speed (10 rpm).
5. Dry the tablet in coating pan with hot air 70 °C for about half an hour after each application.
6. The coating solutions apply with interval of 10—15 minutes each of complete the coating step.

2.	Gelatin	"
3.	D.I Water	"

Procedure:

1. Boil 3.5 L Water in a S.S container.
2. Add Gelatin and Sugar in boil water and stirrer till sugar is completely dissolved.
3. Filter the solution by muslin cloth.
4. Transfer 150 ml of this solution in SS Coating pan and add 60 kg cores.
5. Start the rotation for 15 minutes, maintain the temperature up to 65 °C.
6. Apply 4 to 5 times of 150ml solution of each coating pan.

SUB COATING STAGE

S.NO	Item	Quantity
1.	Sugar	Q.S As per requisition
2.	Talcum	"
3.	Calcium Carbonate	"
4.	D.I Water	"

Procedure:

1. Boil 4 litter D.I water in a neat and clean S.S container.
2. Add sugar in boil water and stirrer till sugar is completely dissolve and then filter the solution.
3. Add Talcum & Calcium Carbonate and mix with a silverson mixer till a smooth syrup obtained.
4. Apply 8 charges of 500 ml sub coating solution in each coating pan and rotate for 10 to15 minutes maintain the temperature at 65 °C.

SUGAR COATING STAGE:

S.NO	Item	Quantity
1.	Sugar	Q.S As per requisition
2.	Food Color	"

3.	D.I Water	"

Procedure:

1. Pour 3 liters water in neat and clean container.
2. Add sugar and food color.
3. Filter the solution with muslin cloth.
4. Apply 10 -12 application of this solution in each coating pan and after each application dry the tablet and check the weight of tablet.

POLISHING STAGE

S.NO	Item	Quantity
1.	Bees Wax	Q.S As per requisition
2.	Carnuba Wax	"
3.	Chloroform	"

Procedure:

1. Dissolve waxes in chloroform.
2. Transfer 100 ml solution to each coating pan.
3. Transfer polished tablet in plastic bucket.
4. Check the weight of tablet.

Prepared By:	Checked By:	Approved By:
…………………….	………………………..	…………………………..

ABC PHARMACEUTICALS
QUALITY CONTROL DEPARTMENT

Work Sheet for Standard Deviation of Coating pan 'Speed Test'

Performance Qualification   **AB-WSVD-01**

Equipment No. : ABPD –T /70 & 70-A to 70-I

Date: ....

S.No.	X	($\underline{X} = \sum X \div n$)	$(X - \underline{X})$	$(X - \underline{X})^2$
1.				
2.				
3.				
4.				
5.				

$\sum X =$			$\sum(X - \overline{X})^2 =$

$$S = \sqrt{\frac{\sum(X - \overline{X})^2}{n-1}}$$

$$S = \sqrt{}$$

$$S = \sqrt{}$$

$S = \div \overline{X} \times 100$

R. S. D = _____ % (Limit: ± 2.0 %)

Remarks: _____

Analyst By: _____ **Q.C.M.** _____

ABC PHARMACEUTICALS
QUALITY CONTROL DEPARTMENT

Work Sheet for Validation of Coating Pan 'Speed Test'   AB-WSVD-055

Date: _____   Performance Qualification

Calibrated Equipment : Stop Watch   Identification No.   ABLI / 73

S #	Time	Equipment No.	Rotation speed 15 ± 2 rpm	Measured rotation speed (rpm)	Acceptable		Analyst
					Yes	No	

ABC PHARMACEUTICALS
QUALITY CONTROL DEPARTMENT

Work Sheet for Validation of Coating Pan 'Blower Temperature test'   AB-WSVD-055

Date: _____   **Performance Qualification**

Calibrated Equipment: Electronic Thermometer with Stainless steel probe.   Identification No. :   ABLI / 51

S#	Time	Equipment No.	Blower Temperature	Observed Blower Temperature	Acceptable	Analyst
1.	9:00	ABPD -T /70	15		Yes	
		ABPD-T /70 - A			Yes	
		ABPD -T /70 - B			Yes	
		ABPD -T /70 - C			Yes	
		ABPD -T /70 - D			Yes	
2.	12:00	ABPD -T /70	15		Yes	
		ABPD -T /70 - A			Yes	
		ABPD -T /70 - B			Yes	
		ABPD -T /70 - C			Yes	
		ABPD -T /70 - D			Yes	
		ABPD -T /70 - I			Yes	

Remarks: _____   Checked By:_____

			(50°C – 70°C)				
					Yes	No	
3.	2:00	ABPD -T /70			Yes		
		ABPD-T /70 - A			Yes		
		ABPD -T /70 - B			Yes		
		ABPD -T /70 - C			Yes		
		ABPD -T /70 - D			Yes		
4.	4:15	ABPD -T /70			Yes		
		ABPD-T /70 - A			Yes		
		ABPD -T /70 - B			Yes		
		ABPD -T /70 - C			Yes		
		ABPD -T /70 - D			Yes		

Remarks: _____   Checked By: _____

ABC PHARMACEUTICALS
QUALITY CONTROL DEPARTMENT

Work Sheet for Validation of Coating Pan 'Blower Air Volume Test'   **AB-WSVD-055**

Date: _____   **Performance Qualification**

Calibrated Equipment: Anemometer   Identification No. :   **ABLI / 63**

S #	Time	Equipment No.	Blower Air Volume (CFM) (800 ---1300)	Observed Blower Air Volume (CFM)	Acceptable		Analyst
					Yes	No	
3.	9:45	ABPD -T /70			Yes		
		ABPD-T /70 - A			Yes		
		ABPD -T /70 - B			Yes		
		ABPD -T /70 - C			Yes		
		ABPD -T /70 - D			Yes		
4.	12:15	ABPD -T /70			Yes		
		ABPD-T /70 - A			Yes		
		ABPD -T /70 - B			Yes		
		ABPD -T /70 - C			Yes		
		ABPD -T /70 - D			Yes		

Remarks: _____   **Checked By:** _____

www.ingramcontent.com/pod-product-compliance
Lightning Source LLC
Chambersburg PA
CBHW061523180526
45171CB00001B/314